Modeling and Control of AC Machine using MATLAB®/SIMULINK

Modeling and Control of AC Machine using MATLAB®/SIMULINK

Mourad Boufadene

CRC Press is an imprint of the
Taylor & Francis Group, an **informa** business

CRC Press
Taylor & Francis Group
6000 Broken Sound Parkway NW, Suite 300
Boca Raton, FL 33487-2742

Printed on acid-free paper
Version Date: 20181129

International Standard Book Number-13: 978-0-367-02302-7 (Hardback)

Library of Congress Cataloging-in-Publication Data

Names: Boufadene, Mourad, author.
Title: Modeling and control of AC machine using MATLAB/SIMULINK / Mourad Boufadene.
Description: Boca Raton : Taylor & Francis, CRC Press, 2019. | Includes bibliographical references.
Identifiers: LCCN 2018044511 | ISBN 9780367023027 (hardback : alk. paper)
Subjects: LCSH: Electric machinery--Alternating current--Computer simulation. | Electric machinery--Alternating current--Automatic control. | MATLAB. | SIMULINK.
Classification: LCC TK2744 .B686 2019 | DDC 621.31/042028553--dc23
LC record available at https://lccn.loc.gov/2018044511

Visit the Taylor & Francis Web site at
http://www.taylorandfrancis.com

and the CRC Press Web site at
http://www.crcpress.com

Contents

Preface

This book introduces electrical machine modeling and control for electrical engineering and science to graduate and undergraduate students, as well as researchers who are working on modeling and control of electrical machines. It targets electrical engineering students who have no time to derive electrical equations for machines like induction motors, and DFIMs especially for those who are working on renewable energy where the dynamical model of doubly fed induction motors is crucial.

It can serve as a text-book for electrical motor modeling-simulation and control; especially modeling of induction motor using different frames of reference. A linear and nonlinear controller is used to enhance the performance of the vector control to induction motor. The reader will find full MATLAB/SIMULINK blocks for simulation and control.

Chapter 1 mathematical modeling of induction motors will be introduced, which are presented in an intuitive and simple manner. The derived dynamical models are represented in different frames dq and $\beta\alpha$, which will be used for control and observation purposes. At the end of this chapter, a full block of MATLAB/SIMULINK is given for the derived dynamical models of induction motors.

Chapter 2 the concept of field-oriented control or vector control will be introduced in an intuitive way, with the application of such technique to doubly fed induction motors using the stator currents and rotor flux as state vector. At the end of this chapter the reader will find MATLAB/SIMULINK codes for whole control strategy which could be used for those who are or are not familiar with MATLAB/SIMULINK.

Chapter 3 a nonlinear controller based on the input-output feedback linearisation technique will be applied to a reduced complex and coupled doubly fed induction motor to enhance the performance of the induction motor at field-oriented control (Stator flux aligning with q axes is zero $\phi_{sq} = 0$), with a linear Luenberger observer which is used to estimate the unmeasured load torque using as a measured state the speed, rotor and stator voltages.

CHAPTER 1

Modeling and Simulation of Induction Motors

CONTROL of an induction motor has been the subject of most research for many years. Recently researchers used the classical control theory to enhance the performance of electrical machines; therefore, the model adopted should be interpreted as accurately as possible so that it captures all basic features of induction motors, which is an essential part for simulation and implementation of observer for state estimation as well as for linear and nonlinear control techniques such us field-oriented control, direct torque control, sliding mode control and feedback linearization control. The induction motor model that can be used for control and observation purposes can be described by a fifth order nonlinear differential equations with stator and rotor voltages (in case of doubly fed induction motor (DFIM)) as inputs, and currents, flux and mechanical speed as state variables.

1.1 INTRODUCTION

The asynchronous motor, often called induction motor, comprises a stator and a rotor which are made of stacked silicon plates and an iron arm respectively, and has slots where the windings are placed. The stator is fixed with windings that are connected to the source,

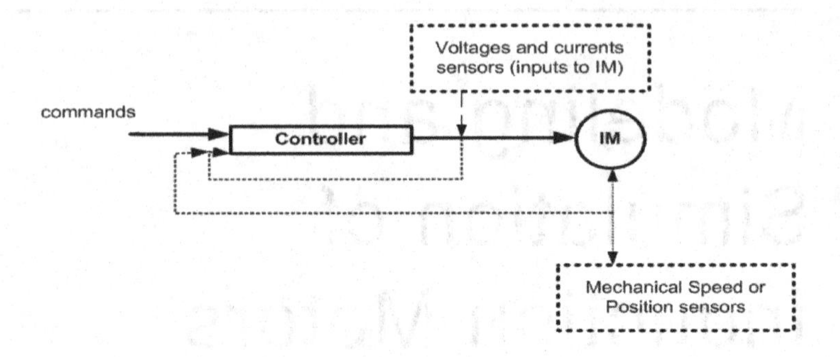

Figure 1.1 Control strategy of induction motor

the aim being to obtain a distribution of magneto-motive forces and a flow as sinusoidal as possible in the gap.

The rotor is mounted on an axis of rotation. Depending on whether the rotor windings are accessible from the outside or are closed on themselves all the time, there are two types of rotors: wound and squirrel cage. Squirrel cage induction motors are used for applications due to cheaper in cost, rugged in construction, and low maintenance. Squirrel cage motors are suitable for applications that required a constant speed, and low starting torque. Wound rotor induction motors are used for applications that required speed control such as generator, elevators, compressors and driving lifts citec5.

Electrical machines are very complex and coupled systems such that it is not possible to find a suitable model that captures all the basic phenomena; therefore, the following simplifying assumptions are used to find a mathematical model:

- The air gap is of uniform thickness, and notching effect is negligible
- The saturation of magnetic circuit, hysteresis and eddy currents are negligible
- The resistances of the windings do not vary with temperature and neglecting the skin effect
- It is recognized that the magneto-motive forces (m.m.f) created by each phase of the two frames is sinusoidal distribution

1.2 MATHEMATICAL MODEL OF INDUCTION MOTOR

The mathematical model of electrical motors in the three phases generally leads to differential equations of higher order, and therefore very complex and complicated systems to be used for control and observation purposes; therefore, the representation of an electrical machine in the two phases becomes crucial. Where the equations of the machine in the three phases are given by:

$$\begin{aligned} V_{sabc} &= R_s I_{sabc} + \dot{\phi}_{sabc} \\ V_{rabc} &= R_r I_{rabc} + \dot{\phi}_{rabc} \end{aligned} \tag{1.1}$$

The vectors $V_{sabc,rabc}$, $I_{sabc,rabc}$, and $\phi sabc, rabc$ are defined by:

$$V_{sabc} = \begin{bmatrix} v_{sa} \\ v_{sb} \\ v_{sc} \end{bmatrix} \quad I_{sabc} = \begin{bmatrix} i_{sa} \\ i_{sb} \\ i_{sc} \end{bmatrix} \quad \phi_{sabc} = \begin{bmatrix} \phi_{sa} \\ \phi_{sb} \\ \phi_{sc} \end{bmatrix} \tag{1.2}$$

$$V_{rabc} = \begin{bmatrix} v_{ra} \\ v_{rb} \\ v_{rc} \end{bmatrix} \quad I_{rabc} = \begin{bmatrix} i_{ra} \\ i_{rb} \\ i_{rc} \end{bmatrix} \quad \phi_{rabc} = \begin{bmatrix} \phi_{ra} \\ \phi_{rb} \\ \phi_{rc} \end{bmatrix} \tag{1.3}$$

Consequently, the equations to be derived using electric theory are based on the transformation of the mathematical model from the thee-to-two-phase model using Clarke and Park transformation techniques. The total flux of the doubly fed induction motor is related to the currents through the following equations:

$$\begin{aligned} \phi_{sabc} &= L_{ss} I_{sabc} + L_{msr} I_{rabc} \\ \phi_{rabc} &= L_{rr} I_{rabc} + L_{mrs} I_{sabc} \end{aligned} \tag{1.4}$$

Where :

$L_{ss}; L_{rr}$: Matrices of self and mutual inductances between stator and rotor phases, respectively

L_{msr} : Matrix of mutual inductances between stator and rotor phases

Since the model given by the three phases is more complicated, a transformation of the system should be used.

1.2.1 Electrical equations of induction motor

The equations derived using electric theory are based on the transformation of the mathematical model from the three- to the two-phase model using the Park transformation technique. Where the equations of the motor with the rotor pulsation $\omega_r = \omega_s - p\Omega$ in the two phases are given by:

$$\begin{aligned} u_{sd} &= R_s i_{sd} + \dot{\phi}_{sd} - \omega_s \phi_{sq} \\ u_{sq} &= R_s i_{sq} + \dot{\phi}_{sq} + \omega_s \phi_{sd} \quad \text{At The stator} \end{aligned} \tag{1.5}$$

$$\begin{aligned} u_{rd} &= R_r i_{rd} + \dot{\phi}_{rd} - \omega_r \phi_{rq} \\ u_{rq} &= R_r i_{rq} + \dot{\phi}_{rq} + \omega_r \phi_{rd} \quad \text{At the rotor} \end{aligned} \tag{1.6}$$

$$\phi_{sd} = L_s i_{sd} + M i_{rd} \tag{1.7}$$
$$\phi_{sq} = L_s i_{sq} + M i_{rq} \text{At The stator} \tag{1.8}$$

$$\phi_{rd} = L_s i_{rd} + M i_{sd} \tag{1.9}$$
$$\phi_{rq} = L_s i_{rq} + M i_{sq} \text{At The rotor} \tag{1.10}$$

The model is represented in a dq frame of reference; this model is used for linear control such as field-oriented control which needs the stator angle; however, for nonlinear control purposes, the most used model is the α ; β model which does not need the estimation of the stator angle; and in order to construct the α; β model just from the dq model, the stator angle has to be set to zero $\theta_s = 0$ and therefore $\omega_s = 0$, hence the rotor pulsation $\omega_r = -p\Omega$. Consequently, the model in fixed frame is given by:

$$\begin{aligned} u_{s\alpha} &= R_s i_{s\alpha} + \dot{\phi}_{s\alpha} \\ u_{s\beta} &= R_s i_{s\beta} + \dot{\phi}_{s\beta} \quad \text{At the stator} \end{aligned} \tag{1.11}$$

$$\begin{aligned} u_{r\alpha} &= R_r i_{r\alpha} + \dot{\phi}_{r\alpha} + p\omega \phi_{rq} \\ u_{r\beta} &= R_r i_{r\beta} + \dot{\phi}_{r\beta} - p\omega \phi_{r\alpha} \quad \text{At the rotor} \end{aligned} \tag{1.12}$$

$$\phi_{s\alpha} = L_s i_{s\alpha} + M i_{r\alpha} \tag{1.13}$$
$$\phi_{s\beta} = L_s i_{s\beta} + M i_{r\beta} \quad \text{At the stator} \tag{1.14}$$

$$\phi_{r\alpha} = L_s i_{r\alpha} + M i_{s\alpha} \tag{1.15}$$
$$\phi_{r\beta} = L_s i_{r\beta} + M i_{s\beta} \quad \text{At the rotor} \tag{1.16}$$

Remark 1 *In order to construct the induction motor model from the previous dynamical equations, the rotor voltage has to be set to zero $u_r = 0$; then you will get the induction machine model.*

1.2.2 Mechanical equations of induction motor

The mathematical model of the induction motors must contain the mechanical speed and the electromagnetic torque equations which are the most important equations for control purposes. The electromagnetic torque can be derived from power expression as:

$$P_e = pM\Omega \times Im([i_s^* \times i_s]) \tag{1.17}$$

Then the electromagnetic torque is given by:

$$C_e = \frac{P_e}{\Omega} = pM \times Im([i_s^* \times i_s]) \tag{1.18}$$

Where p is the number of pair pole; therefore, the electromagnetic torque equations are written in many forms as:

$$C_e = pM(\phi_{sd} i_{sq} - \phi_{sq} i_{sd}) \tag{1.19}$$

$$C_e = pM(\phi_{rq} i_{rd} - \phi_{rd} i_{rq})$$

$$C_e = \frac{pM}{L_r}(\phi_{rd} i_{sq} - \phi_{rq} i_{sd}) \tag{1.20}$$

$$C_e = \frac{pM}{L_r}(\phi_{sq} i_{rd} - \phi_{sd} i_{rq})$$

The mechanical speed is deduced from the fundamental law of general mechanics as follows:

$$J_m \dot{\Omega} = C_e - f_m \Omega - \tau_l \tag{1.21}$$

1.3 DYNAMICAL MODEL OF DOUBLY FED INDUCTION MOTOR

The model of the induction motor is represented in state space model using as state variables the stator currents i_{sd}, i_{sq} and rotor flux ϕ_{rd}, ϕ_{rq}; and in order to do so, the rotor currents as well as the stator flux equations have to be eliminated from Equations (1.6) and (1.7); therefore the total flux equations are used to find the rotor currents and stator flux states as:

$$i_{rd} = \frac{\phi_{rd} - M i_{sd}}{L_r}$$
$$i_{rq} = \frac{\phi_{rq} - M i_{sq}}{L_r} \tag{1.22}$$

$$\phi_{sd} = (L_s - \frac{M^2}{L_r}) i_{sd} + \frac{M}{L_r}\phi_{rd}$$
$$\phi_{sq} = (L_s - \frac{M^2}{L_r}) i_{sq} + \frac{M}{L_r}\phi_{rq} \tag{1.23}$$

We put $T_r = \frac{L_r}{R_r}$, $\sigma = (L_s - \frac{M^2}{L_s L_r})$, and $K = \frac{M}{L_s L_r \sigma}$ equations (1.23) will be:

$$\phi_{sd} = L_s \sigma i_{sd} + \frac{M}{L_r}\phi_{rd}$$
$$\phi_{sq} = L_s \sigma i_{sq} + \frac{M}{L_r}\phi_{rq} \tag{1.24}$$

1.3.1 State space model of DFIM in dq

The dynamical model of the doubly fed induction motors is derived by substituting Equations (1.22) and (1.24) into Equations (1.6) and (1.7) gives the following state space DFIM:

$$\dot{i}_{sd} = -\gamma i_{sd} - \omega_s i_{sq} + \frac{K}{T_r}\phi_{rd} - K p \Omega \phi_{rq} + \frac{1}{\sigma L_s} u_{sd} - K u_{rd}$$
$$\dot{i}_{sq} = -\gamma i_{sq} + \omega_s i_{sd} + \frac{K}{T_r}\phi_{rq} + K p \Omega \phi_{rd} + \frac{1}{\sigma L_s} u_{sq} - K u_{rq}$$
$$\dot{\phi}_{rd} = \frac{M}{T_r} i_{sd} - \frac{1}{T_r}\phi_{rd} + \omega_r \phi_{rq} + u_{rd} \tag{1.25}$$
$$\dot{\phi}_{rq} = \frac{M}{T_r} i_{sq} - \frac{1}{T_r}\phi_{rq} - \omega_r \phi_{rd} + u_{rq}$$

1.3.2 State space model of DFIM in $\alpha\beta$

In order to write the state space of the DFIM in a stationary frame of reference (α, β),

$$i_{r\beta} = \frac{\phi_{r\alpha} - M i_{s\alpha}}{L_r} \tag{1.26}$$
$$i_{r\alpha} = \frac{\phi_{r\beta} - M i_{s\beta}}{L_r}$$

$$\begin{aligned}\phi_{s\alpha} &= L_s\sigma i_{s\beta} + \frac{M}{L_r}\phi_{r\alpha} \\ \phi_{s\beta} &= L_s\sigma i_{s\beta} + \frac{M}{L_r}\phi_{r\beta}\end{aligned} \tag{1.27}$$

put the stator pulsation $\omega_s = 0$ and substitute Equations (1.26) and (1.27) into Equations (1.11) and (1.12).

$$\begin{aligned}\dot{i}_{s\beta} &= -\gamma i_{s\beta} + \frac{K}{T_r}\phi_{r\alpha} + Kp\Omega\phi_{r\beta} + \frac{1}{\sigma L_s}u_{s\beta} - Ku_{r\beta} \\ \dot{i}_{s\alpha} &= -\gamma i_{s\alpha} + \frac{K}{T_r}\phi_{r\beta} - Kp\Omega\phi_{r\alpha} + \frac{1}{\sigma L_s}u_{s\alpha} - Ku_{r\alpha} \\ \dot{\phi}_{r\alpha} &= \frac{M}{T_r}i_{s\alpha} - \frac{1}{T_r}\phi_{r\alpha} - p\Omega\phi_{r\beta} + u_{r\alpha} \\ \dot{\phi}_{r\beta} &= \frac{M}{T_r}i_{s\beta} - \frac{1}{T_r}\phi_{r\beta} + p\Omega\phi_{r\alpha} + u_{r\beta}\end{aligned} \tag{1.28}$$

Remark 2 *In order to construct the induction motor model from the previous dynamical equations, the rotor voltage has to be set to zero $u_r = 0$; then you will simply get the induction motor (IM) model.*

1.4 DYNAMICAL MODEL OF INDUCTION MOTORS

In order to construct the induction machine model from the previous dynamical equations, the rotor voltages has to be set to zero $u_r = 0$; then you will simply get the induction machine (IM) model. Therefore, Equations (1.29) and (1.30) will be:

$$\begin{aligned}u_{sd} &= R_s i_{sd} + \dot{\phi}_{sd} - \omega_s\phi_{sq} \\ u_{sq} &= R_s i_{sq} + \dot{\phi}_{sq} + \omega_s\phi_{sd} \quad \text{At the stator}\end{aligned} \tag{1.29}$$

$$\begin{aligned}0 &= R_r i_{rd} + \dot{\phi}_{rd} - \omega_r\phi_{rq} \\ 0 &= R_r i_{rq} + \dot{\phi}_{rq} + \omega_r\phi_{rd} \quad \text{At the rotor}\end{aligned} \tag{1.30}$$

1.4.1 State space model of induction motor in dq

The model of the induction motor is represented in the state space model using as state variables the stator currents i_{sd}, i_{sq} and rotor flux ϕ_{rd}, ϕ_{rq}; and in order to do so, the rotor currents as well as the stator flux equations have to be eliminated from Equations (1.6) and (1.7); therefore, the total flux Equations (1.22) and (1.24) are substituted into (1.29, 1.30); the asynchronous motor mode is given by:

$$
\begin{aligned}
\dot{i}_{sd} &= -\gamma i_{sd} - \omega_s i_{sq} + \frac{K}{T_r}\phi_{rd} + Kp\Omega\phi_{rq} + \frac{1}{\sigma L_s}u_{sd} \\
\dot{i}_{sq} &= -\gamma i_{sq} + \omega_s i_{sd} + \frac{K}{T_r}\phi_{rq} - Kp\Omega\phi_{rd} + \frac{1}{\sigma L_s}u_{sq} \\
\dot{\phi}_{rd} &= \frac{M}{T_r}i_{sd} - \frac{1}{T_r}\phi_{rd} - p\Omega\phi_{rq} \\
\dot{\phi}_{rq} &= \frac{M}{T_r}i_{sq} - \frac{1}{T_r}\phi_{rq} + p\Omega\phi_{rd}
\end{aligned}
\tag{1.31}
$$

1.4.2 State space model of induction motor in $\alpha\beta$

The $\alpha\beta$ induction motor model can be derived from the dq model by setting stator pulsation $\omega_s = 0$ in Equations (1.29) and (1.30), and yields:

$$
\begin{aligned}
\dot{i}_{s\beta} &= -\gamma i_{s\beta} + \frac{K}{T_r}\phi_{r\alpha} + Kp\Omega\phi_{r\beta} + \frac{1}{\sigma L_s}u_{s\beta} \\
\dot{i}_{s\alpha} &= -\gamma i_{s\alpha} + \frac{K}{T_r}\phi_{r\beta} - Kp\Omega\phi_{r\alpha} + \frac{1}{\sigma L_s}u_{s\alpha} \\
\dot{\phi}_{r\alpha} &= \frac{M}{T_r}i_{s\alpha} - \frac{1}{T_r}\phi_{r\alpha} - p\Omega\phi_{r\beta} \\
\dot{\phi}_{r\beta} &= \frac{M}{T_r}i_{s\beta} - \frac{1}{T_r}\phi_{r\beta} + p\Omega\phi_{r\alpha}
\end{aligned}
\tag{1.32}
$$

1.5 MATLAB/SIMULINK MODEL OF DFIM

For control and observation purposes, the dynamical model of the doubly fed induction machine should be represented in a continuous state space form as follows:

$$
\dot{x} = f(x) + gu \tag{1.33}
$$

Where:

$$
f(x) = \begin{bmatrix}
-\gamma i_{s\beta} + \frac{K}{T_r}\phi_{r\alpha} + Kp\Omega\phi_{r\beta} \\
-\gamma i_{s\alpha} + \frac{K}{T_r}\phi_{r\beta} - Kp\Omega\phi_{r\alpha} \\
\frac{1}{T_s}i_{s\beta} - \frac{1}{T_s}\phi_{r\beta} - p\Omega\phi_{r\beta} \\
\frac{M}{T_r}i_{s\beta} - \frac{1}{T_r}\phi_{r\beta} + p\Omega\phi_{r\alpha}
\end{bmatrix}
\tag{1.34}
$$

$$g(x) = \begin{bmatrix} 0 & \frac{1}{\sigma L_s} & 0 & -K \\ \frac{1}{\sigma L_s} & 0 & -K & 0 \\ 0 & 0 & 1 & 0 \\ 0 & 0 & 0 & 1 \end{bmatrix} \begin{bmatrix} u_{s\alpha} \\ us\beta \\ u_{r\alpha} \\ u_{r\beta} \end{bmatrix} \tag{1.35}$$

The validation of a dynamical model is an essential part for control purposes; therefore, the MATLAB/SIMULINK software has been used to validate the doubly fed induction motor. A MATLAB S-function is used to perform the required simulation; In order to use the S-function in simulation the following points have to be known:

- The number of continuous states
- The number of inputs
- The number of outputs

```
sizes=simsizes;
sizes.NumContStates=5;
sizes.NumDiscStates=0;
sizes.NumOutputs=6;
sizes.NumInputs=6;
sizes.DirFeedthrough=0;
sizes.NumSampleTimes=1;
sys=simsizes(sizes);
str=[];
ts=[0 0];
x0=[0 0 0 0 0];   % intial conditions
```

According to the dynamical model of Equations (1.21), (1.25) and (1.35) , the number of stats are 5 in dq frame of reference, which are written in MATLAB as follows:

```
case 1
sys(1)=-gamma*x(1)+(K/Tr)*x(3)-u(6)*x(2)
        -K*p*x(5)*x(4)+(1/(Ls*sigma))*u(1)-K*u(3);
sys(2)=-gamma*x(2)+(K/Tr)*x(4)+u(6)*x(1)
        +K*p*x(5)*x(3)+(1/(Ls*sigma))*u(2)-K*u(4);
sys(3)=(M/Tr)*x(1)-(1/Tr)*x(3)+(u(6)-p*x(5))*x(4)+u(3);
sys(4)=(M/Tr)*x(2)-(1/Tr)*x(4)-(u(6)-p*x(5))*x(3)+u(4);
sys(5)=((p*M)/(Jm*Ls))*(x(3)*x(2)-x(1)*x(4))-u(5)/Jm
       -(fm/Jm)*x(5);
```

The number of inputs is represented by the stator and rotor voltages, respectively, as well as the load torque of the machine which is considered as a perturbation to the machine. The number of outputs is written as:

```
case 3
```

```
sys(1)=x(1);                          % The isd current
sys(2)=x(2);                          % The isq current
sys(3)=x(3);                          % The frd flux
sys(4)=x(4);                          % The frq flux
sys(5)=x(5);                          % The mechanical speed
sys(6)=mm*(-x(3)*x(2)+x(4)*x(1));     % The machine torque C_e
```

The derived equations are based on the transformation of the mathematical model of the three-to-two-phase models using Park transformation techniques represented by the MATLAB code shown in appendix A; the following figures are obtained using the machine parameters in appendix B. The MATLAB/SIMULINK block of the doubly fed induction motor is shown on the following figures: Inside the Figure 1.2 block DFIM there will be the

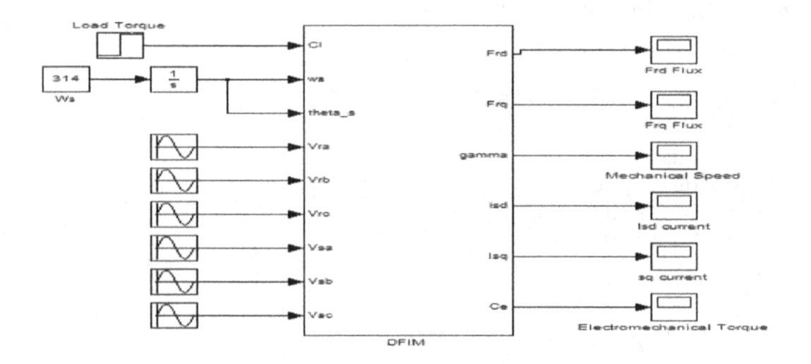

Figure 1.2 MATLAB/SIMULINK Block of the doubly fed induction motor

S-function that describes the model of the doubly fed induction motor as well as the transformation blocks shown in Figure 2.1; hence the following figure shows the inside DFIM block: The mechanical speed and electromechanical torque using a load torque of $5N.m$ at $t = 2s$ is shown in Figure 2.2:

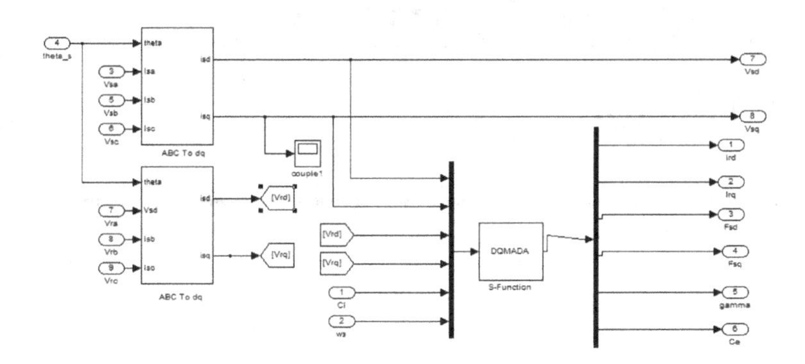

Figure 1.3 MATLAB/SIMULINK Block of DFIM

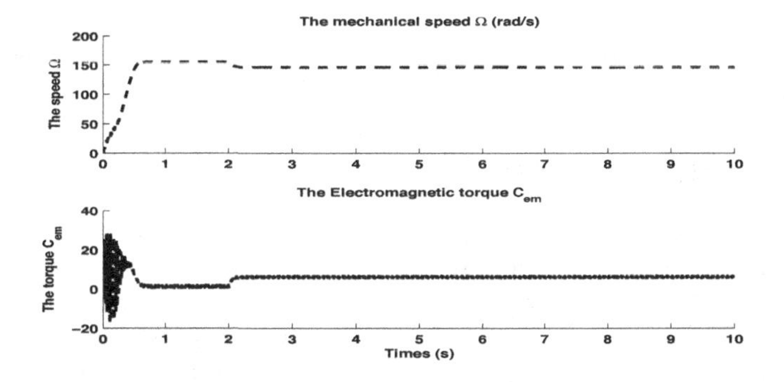

Figure 1.4 The speed and electromagnetic torque

CHAPTER 2

Vector Control of DFIM

LINEAR control techniques can be used to control linear systems, and since the induction machine has a complex and coupled nonlinear dynamical mathematical model, therefore linear control techniques using merely Proportional Integrator Derivative (PID) controller cannot be applied directly to the complex model unless its reduced to a simple and linear model.

In this case, linearization, is achieved using field-oriented techniques which can make the highly complex model of the motor to be controlled as a linear direct current motor. The electromagnetic torque of an alternating current (AC) motor can be controlled from the stator or rotor currents which allows a very fast response of the torque.

2.1 INTRODUCTION

The control of an AC machine is generally carried out by two technique a classical techniques such as (scalar control, vector control, and direct torque control) and modern control techniques such as nonlinear controllers. Hence, for applications requiring high dynamic performance, the electromagnetic torque plays an important role which is to achieve better dynamic performance.

The current of the motor or the armature magnetomotive foce is set at right angles to the axis of the flux excitation. Regardless of the mechanical speed and the load torque, the electromagnetic torque is proportional to the product of the flux excitation and the armature current. If the machine is excited separately and is maintained at a constant excitation flux, the torque is directly proportional to the armature current; therefore, leads to a good dynamic performance since the torque is controlled as fast as the aramtrue current can be [2, 3, 10]. On the other side, in an asynchronous machine, the angle between the rotating field of the stator and the rotor varies with the load torque, which results in a complex interaction and dynamic oscillatory response.

For a situation equivalent to that of the current machine, we introduce the technique of vector control. The principle of it is to replace the model of the asynchronous motor with a structure similar to the direct current motor.

Indeed, in the absence of saturation and armature reaction, the main flow of the inductor is proportional to the excitation current. It is not affected by the armature current due to the perpendicular orientation of the stator or rotor flux. Therefore, the electromagnetic torque of a Direct Current (DC) motor with separate excitation is directly proportional to the armature current for a constant flow, which has a quick adjustment of the torque. And for the asynchronous motor, the angle between the stator and rotor fields is different from $\frac{\pi}{2}$. The idea proposed by Blaschke and Hasse is to decompose the vector of the stator or rotor currents into two components: one produces flow, and the other produces the torque. This allows for an angle of $\frac{\pi}{2}$ between the two flows of the machine, and we obtain a characteristic similar to machine that is separately excited.

2.2 FLUX ORIENTATION OF AC MACHINE

The field-oriented control is based on the choice of the frame of reference in which the axes are oriented, which are the stator or rotor flux orientations respectively. The orientations of the rotor flux or stator flux on the dq axes are the alignments of the rotor flux or the stator flux on the d axes as described by the following expression:

$$\begin{aligned} \phi_{rd} &= \phi_r \quad \phi_{rq} = 0 \\ \phi_{sd} &= \phi_s \quad \phi_{sq} = 0 \end{aligned} \tag{2.1}$$

Where the expression of electromagnetic torque of the AC machine in terms of the stator currents and rotor flux is given by:

$$C_e = \frac{pM}{L_r}(\phi_{rd} i_{sq} - \phi_{rq} i_{sd})$$
$$C_e = \frac{pM}{L_s}(\phi_{sq} i_{rd} - \phi_{sd} i_{rq}) \tag{2.2}$$

After having defined the axes in which the orientation will be; the electromagnetic torque equation will be:

$$C_e = \frac{pM}{L_r}\phi_{rd} i_{sq}$$
$$C_e = -\frac{pM}{L_s}\phi_{sd} i_{rq} \tag{2.3}$$

And since that the flux is constant, the torque equation is controlled from the stator or rotor currents, so the torque can take the form of a Direct Current (DC) motor.

2.3 FIELD-ORIENTED CONTROL OF DFIM

Under the simplification assumptions and balanced condition, the equivalent two-phase model of doubly fed induction motor in the stator (d, q) fixed reference frame related to the stator can be obtained. So the model can be written in a compact form as:

$$\dot{x} = f(x) + gu \tag{2.4}$$

where the state vector x is defined as:

$$x = [i_{sd}, i_{sq}, \phi_{rd}, \phi_{rq}, \Omega]^T \tag{2.5}$$

and the input vector is:

$$u = [u_{sd}, \ u_{sq}]^T \tag{2.6}$$

with

$$f(x) = \begin{bmatrix} -\gamma i_{sd} + \omega_r i_{sq} + \frac{K}{T_r \phi_{rd} - K\omega\phi_{rq} - K u_{rd}} \\ -\gamma i_{sq} - \omega_r i_{rd} + \frac{K}{T_r}\phi_{rq} + K\omega\phi_{rd} - K u_{rq} \\ \frac{M}{T_r} i_{sd} - \frac{1}{T_r}\phi_{rd} + \omega_s \phi_{rq} + u_{rd} \\ \frac{M}{T_r} i_{sq} - \frac{1}{T_r}\phi_{rq} - \omega_s \phi_{rd} + u_{rq} \\ \frac{pM}{J_m L_r}(\phi_{rd} i_{sq} - \phi_{rq} i_{sd}) - \frac{f_m}{J_m}\Omega - \frac{1}{J_m} C_r \end{bmatrix} \tag{2.7}$$

$$g = \begin{bmatrix} \frac{1}{\sigma L_s} & 0 & 0 & 0 \\ 0 & \frac{1}{\sigma L_s} & 0 & 0 \end{bmatrix}^T . \tag{2.8}$$

where the parameters σ, γ, K, T_s, T_r are defined as follows

$$\begin{array}{ll} \sigma = 1 - \frac{M^2}{L_r L_s}, & \gamma = \frac{1}{T_s} + \frac{1-\sigma}{T_s}\frac{1}{\sigma} \\ K = \frac{1-\sigma}{\sigma}, & T_s = \frac{L_s}{R_s}, \; T_r = \frac{L_r}{R_r} \end{array} \tag{2.9}$$

σ is the scattering coefficient, T_r, T_s are the time constant of the rotor and stator dynamics, J_m is the rotor inertia, f_m is the mechanical viscous damping, p is the number of pole pairs, and c_r is the external load torque. The state variables i_{sd}, i_{sq}, ϕ_{rd}, ϕ_{rd}, ϕ_{rq}, u_{sd}, u_{sq}, u_{rd}, u_{rq} are the stator currents, rotor flux linkages, stator terminal voltage, rotor terminal voltage, respectively, and L_r, L_s, M, R_r, R_s are rotor inductance, stator inductance, mutual inductance, stator resistance and rotor resistance, respectively.

Remark 3 *The field-oriented control of induction machine can be derived from the previous model by setting $u_r = 0$ and $\omega = -\omega$, or use any model that has been derived from the previous chapter for induction motor (IM) and follow the same steps listed in the subsequent sections.*

2.3.1 Rotor flux orientation control

Oriented vector control of rotor flux is frequently used because it eliminates the influence of the leakage reactance rotor and stator and gives better results than methods based on the orientation of the stator flux or airgap [?, 5]. This control is achieved by orienting the rotor flux following the direct axis d of the rotating frame as shown in Fig. 2.1, so

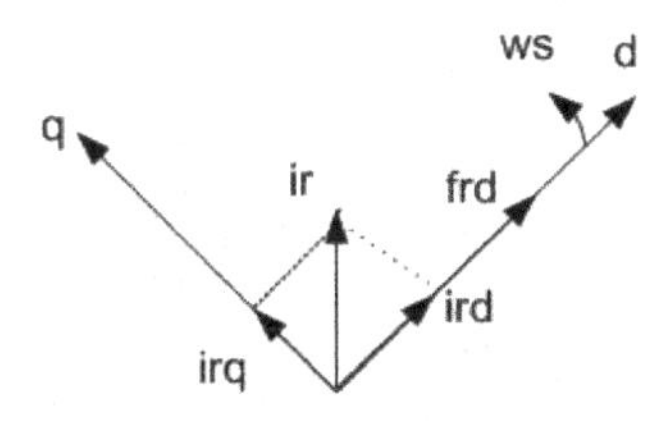

Figure 2.1 The orientation of stator flux

$$\begin{cases} \phi_{rd} & = \phi_r \\ \phi_{rq} & = 0 \end{cases} \tag{2.10}$$

So by introducing Equation (2.10) into Equation (2.7), we get:

$$u_{sd} = R_s i_{sd} + L_s \sigma \frac{di_{sd}}{dt} + \frac{M d\phi_r}{L_r dt} - \omega_s L_s \sigma i_{sq} \tag{2.11}$$

$$u_{sq} = R_s i_{sq} + L_s \sigma \frac{di_{sq}}{dt} + \omega_s \frac{M \phi_r}{L_r} + \omega_s L_s \sigma i_{sd} \tag{2.12}$$

with a rotor flux and the rotor angle estimation written as follows:

$$T_r \frac{d\phi_r}{dt} + \phi_r = M i_{sd} + T_r v_{rd} \tag{2.13}$$

$$\omega_r = \frac{M i_{sq} + T_r v_{rq}}{T_r \phi_r} \tag{2.14}$$

The electromagnetic torque will be reduced to:

$$C_e = \frac{pM}{L_r} \phi_r i_{sq} \tag{2.15}$$

The proportional integral (PI) controller is used to control the current vector, but this controller can only control a linear system, so equations (2.11) and (2.12) must be linearized first by the decoupling

$$u_{sd} = v_{sd} + e_d \tag{2.16}$$

$$u_{sq} = v_{sq} + e_q \tag{2.17}$$

where:

$$v_{sd} = R_s i_{sd} + L_s \sigma \frac{di_{sd}}{dt} \tag{2.18}$$

$$v_{sq} = R_s i_{sq} + L_s \sigma \frac{di_{sq}}{dt} \tag{2.19}$$

$$e_d = \frac{M d\phi_r}{L_r dt} - \omega_s L_s \sigma i_{sq} \tag{2.20}$$

$$e_q = \omega_s \frac{M \phi_r}{L_r} + \omega_s L_s \sigma i_{sd} \tag{2.21}$$

Where :

e_d, e_q: represent the electromotive forces compensation

v_{sd}, v_{sq}: represent the emf of compensation that allow decoupling of the control current i_{sd} and current i_{sq}.

Where by introducing Laplace transformation to Equations (2.18) and (2.19), so that the model that we will be used for compensation is shown in Figure (2.2)

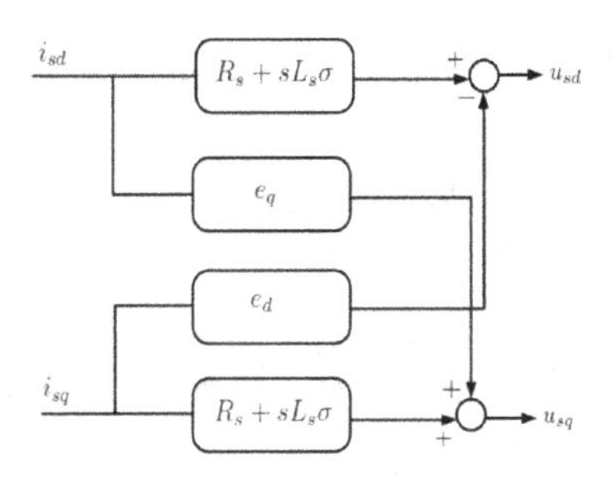

Figure 2.2 Compensation scheme

2.3.2 Controller tuning parameters

2.3.2.1 Current controller tuning parameters

Assuming that the decoupling is achieved, we will get:

$$\frac{i_{sd}}{v_{sd}} = \frac{1}{R_s + sL_s\sigma} \tag{2.22}$$

The loop current i_{sd} is represented by the block diagram of Fig (2.3):

where $F(s)$ and PI are

$$F_d(s) = \frac{1}{R_s + sL_s\sigma} \tag{2.23}$$

$$PI = K_p + \frac{K_i}{s} \tag{2.24}$$

The transfer function in the closed loop is as follows.

$$\frac{i_{sd}}{i_{sd_{ref}}} = \frac{K_p s + K_i}{L_s\sigma s^2 + (R_s + K_p)s + K_i} \tag{2.25}$$

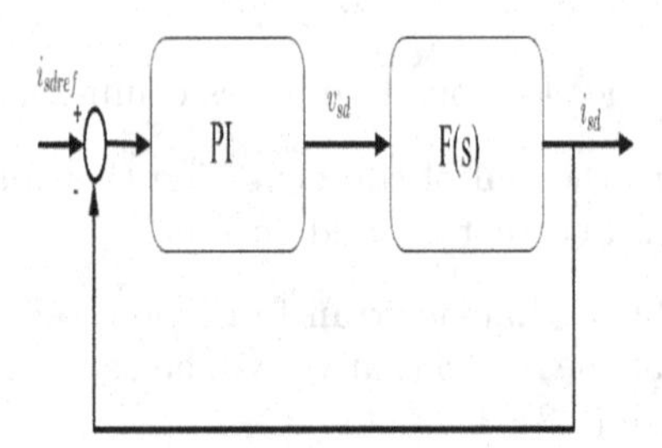

Figure 2.3 Block diagram of i_{sd} current control

The system in the closed loop may have behaved like a first order system whose transfer function is $\frac{1}{\tau s+1}$, such that

$$\frac{K_p s + K_i}{L_s \sigma s^2 + (R_s + K_p)s + K_i} = \frac{1}{\tau s + 1} \tag{2.26}$$

After simplification we have:

$$\begin{cases} K_p = & \frac{L_s \sigma}{\tau} \\ \\ K_i = & \frac{R_s}{\tau} \end{cases} \tag{2.27}$$

with $\tau < \tau_e$ where τ_e is the electrical time constant of the system, which is in our case $\tau_e = \frac{L_s \sigma}{R_s}$, as well as for the i_{sq} parameter.

2.3.2.2 Speed controller tuning parameters

The speed controller is used to determine the reference torque, in order to maintain the corresponding speed. The dynamics of the speed is given by the mechanical equation

$$C_e - C_r = J_m \frac{d\Omega}{dt} + f_m \Omega \tag{2.28}$$

using Laplace transformation we find that :

$$\Omega = \frac{C_e - C_r}{J_m s + f_m} \tag{2.29}$$

The simplified block diagram of the control system with PI controller is given in Fig (2.4)

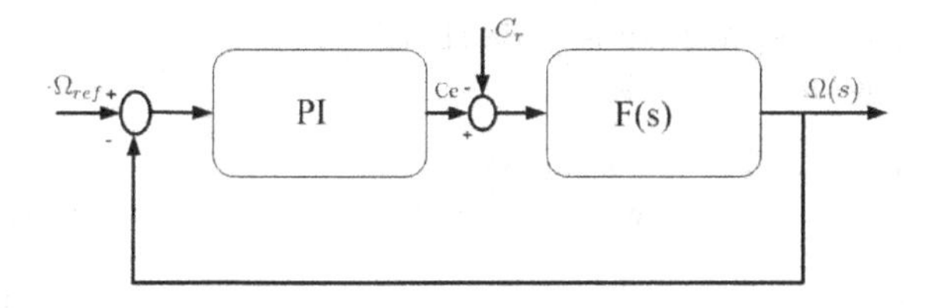

Figure 2.4 Block diagram of speed control

where $F(s) = \frac{1}{J_m s + f_m}$ and The transfer function in an open loop is:

$$\Omega(s) = \frac{K_p s + K_i}{J_m s^2 + f_m s} \tag{2.30}$$

The transfer function in a closed loop is:

$$\begin{aligned} \Omega(s) &= \frac{K_p s + K_i}{J_m s^2 + (f_m + K_p)s + K_i} \Omega_{ref} \\ &- \frac{K_p s + K_i}{J_m s^2 + (f_m + K_p)s + K_i} C_r \end{aligned} \tag{2.31}$$

The perturbation has to be eliminated by putting $c_r = 0$ then by identification to a second order characteristic equation, we have:

$$\begin{aligned} K_p &= 2\zeta\omega_n J_m - f_m \\ K_i &= \omega_n^2 J_m \end{aligned} \tag{2.32}$$

2.4 MATLAB/SIMULINK BLOCK OF THE FIELD-ORIENTED CONTROL

The field oriented control SIMULINK block of the DFIM contains two main blocks:

- The field weakening block, which is used to control the flux of the machine using the desired mechanical speed Ω_{ref}
- Four PI controllers that control the speed of the machine from the stator voltages side

2.4.1 The field weakening MATLAB/SIMULINK block

The field weakening block allows to control the flux using the desired mechanical speed which is defined by the following Equation:

$$\phi_r = \phi_{ref} \tag{2.33}$$

$$\phi_r = \phi_r \frac{\Omega_{ref}}{\Omega} \tag{2.34}$$

The MATLAB/SIMULINK block is shown in Fig 2.5

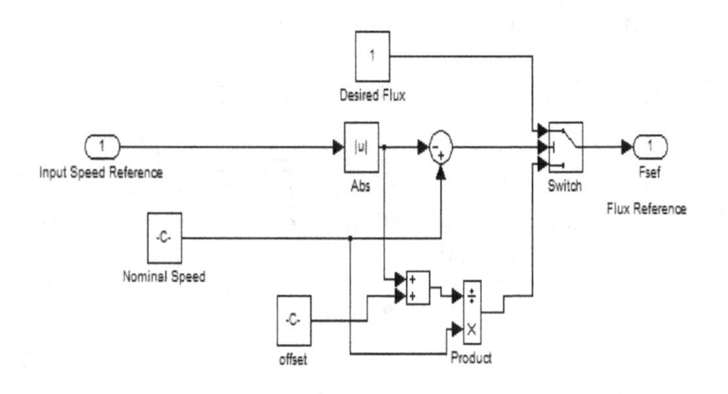

Figure 2.5 Field weakening block of DFIM

2.4.2 The field-oriented MATLAB/SIMULINK block

The complete MATLAB/SIMULINK block of the filed-oriented control of doubly fed induction motor is shown in Fig 2.6

2.5 FIELD ORIENTED CONTROL PERFORMANCE

Controlling the AC machine from the stator voltage side yields to better performance of the machine without including any variations like parameters or load torque; on the other hand, the speed of the

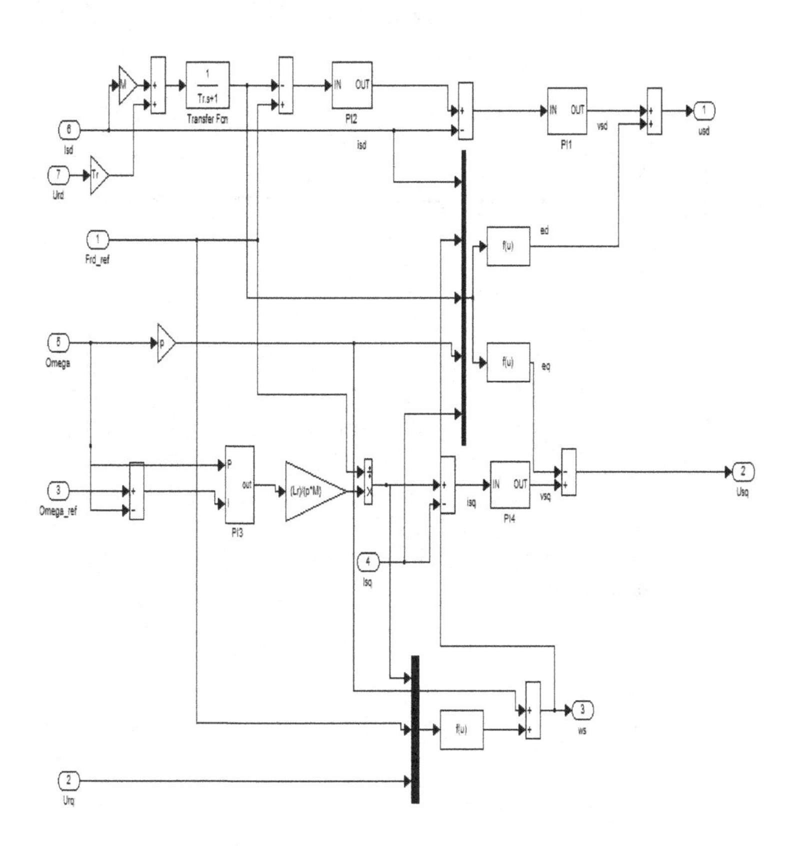

Figure 2.6 Field-oriented control block of DFIM

machine is affected by the application of a load torque which could be compensated for using nonlinear controllers.

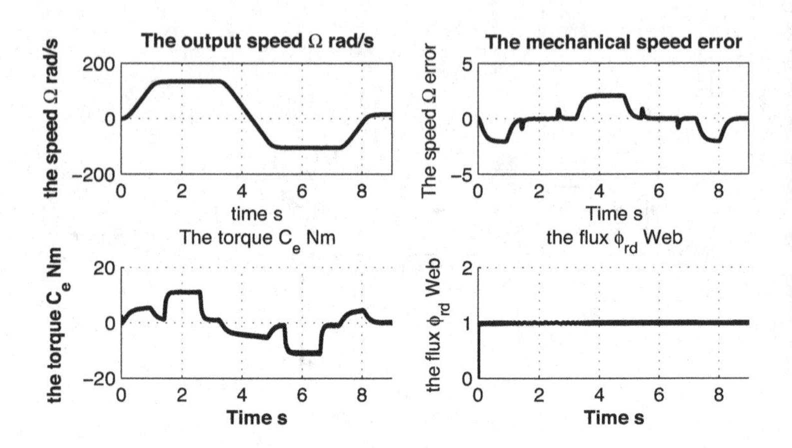

Figure 2.7 FOC performance

However, controlling the DFIM from the rotor side yields to undesirable output oscillations for the mechanical speed. Thus nonlinear control techniques such as sliding mode and feedback linearization controllers to the speed and flux as output vector can be used to overcome this kind of problem applied to a special dynamical model of DFIM.

CHAPTER 3

Vector Control Using Nonlinear Controller for DFIM

THE doubly fed induction motor can be controlled from either a stator or a rotor voltage sid;, however, controlling the motor from the rotor voltage side using field-oriented control leads to undesirable oscillations at the output of the mechanical speed even by increasing the PI controller gains. Thus nonlinear controllers such us feedback linearization and sliding mode control can be applied as a solution to overcome this problem.

A robust nonlinear controller is applied to a reduced dynamical model of doubly fed induction motor, which is based on the input-output feedback linearization technique to enhance the performance of the motor, and since the load torque is unmeasured and has to be compensated using observers. Thus a linear Leunberger observer is applied using as measured states he speed , stator and rotor voltages.

3.1 INTRODUCTION

The doubly fed induction motor (DFIM) has been known since 1899 as a wound rotor asynchronous machine that is supplied from the stator and rotor sides with an external voltage source [12, 28]. The DFIM is highly nonlinear and coupled, multivariable system that required more complex methods to be controlled, hence DFIM constitutes a theoretically challenging control problem [18]. The wound rotor doubly fed asynchronous machine has been the subject of most research primarily for its operation as a generator in applications of wind energy. This work involves the operation in variable speed motor, for improving the robustness of the control of the DFIM.

The DFIM drive wound rotor has the advantage that it can be controlled from the stator or rotor as well as from both (stator and rotor) by various possible combinations compared to the squirrel-cage machine. The field oriented control technique that has been established by [8] is the most important development in control area. However, this technique is affected by unknown disturbances; therefore, extensive work has been established to find a new solution such as sliding mode control ,feedback linearization control and predictive control [9, 10, 19] to improve the dynamic response and reduce the complexity of the field-oriented control. In this chapter, a Multi-input Multi-output robust nonlinear controller [29] based on the technique of linearization to adjust the speed of the DFIM is introduced and applied with a load torque observer in order to estimate the unmeasured load disturbance. This technique made the linearized system in the canonical form where the pole placement technique is applied to find the controller parameters.

3.2 FIELD-ORIENTED CONTROL ENHANCEMENT USING NONLINEAR CONTROL

Feedback linearization uses mainly a state transformation that enable us to express the system in a new coordinate system that makes it linear, so that a state linearization in the new coordinates is achieved. Theoretical background and procedures using MATLAB symbolic computation for nonlinear controllers such as feedback linearization and sliding mode controller is found in [29].

3.2.1 The mathematical model of doubly fed induction motor

The doubly fed induction motor model can be described in the synchronous frame with the stator flux aligned on the d axis such that $\phi_{sq} = 0$, by the following state equations:

$$\dot{x} = f(x) + gu + B\tau_l \tag{3.1}$$

where the state vector x and the input vector u are defined as:

$$x = \left[i_{rd}, i_{rq}, \phi_{sd}, \Omega\right]^T; \quad u = \left[u_{rd}, u_{rq}\right]^T \tag{3.2}$$

The stator pulsation Equation (3.4) has been substituted onto the dynamical model, therefore the function $f(x)$ is defined as:

$$f(x) = \begin{bmatrix} -\gamma i_{rd} + \frac{K}{T_s}\phi_{sd} + \frac{M i_{rq}^2}{T_s \phi_{sd}} - p\Omega i_{rq} + \frac{i_{rq}}{\phi_{sd}} u_{sq} - K u_{sd} \\ -\gamma i_{rq} - \frac{M i_{rq} i_{rd}}{T_s \phi_{sd}} + p\Omega i_{rd} - \frac{i_{rd}}{\phi_{sd}} u_{sq} + K p \Omega \phi_{sd} - K u_{sq} \\ \frac{M}{T_s} i_{rd} - \frac{\phi_{sd}}{T_s} + u_{sd} \\ -\frac{pM}{J_m L_r}\phi_{sd} i_{rd} - \frac{f_m}{J_m}\Omega \end{bmatrix} \tag{3.3}$$

Where the stator pulsation equation is given by:

$$\omega_s = \frac{\frac{M}{T_s} i_{rq} + u_{sq}}{\phi_{sd}} \tag{3.4}$$

The control input matrices are defined as:

$$g = \begin{bmatrix} \frac{1}{\sigma L_r} & 0 & 0 & 0 \\ 0 & \frac{1}{\sigma L_r} & 0 & 0 \end{bmatrix}^T; \quad B = \begin{bmatrix} 0 & 0 & 0 & -\frac{1}{J_m} \end{bmatrix}^T \tag{3.5}$$

The parameters σ, γ, K, T_s are given by:

$$\begin{array}{ll} \sigma = 1 - \frac{M^2}{L_r L_s}, & \gamma = \frac{1}{T_s} + \frac{1-\sigma}{T_s}\frac{1}{\sigma} \\ K = \frac{1-\sigma}{\sigma}, & T_s = \frac{L_s}{R_s},\ T_r = \frac{L_r}{R_r} \end{array} \tag{3.6}$$

Simulink block for the open loop model

The model of Equation (3.3) is a modified model that will be used to construct the nonlinear controller, whereas a completed model of the induction machine is used to describe the machine for control purposes which are given by the following MATLAB code:

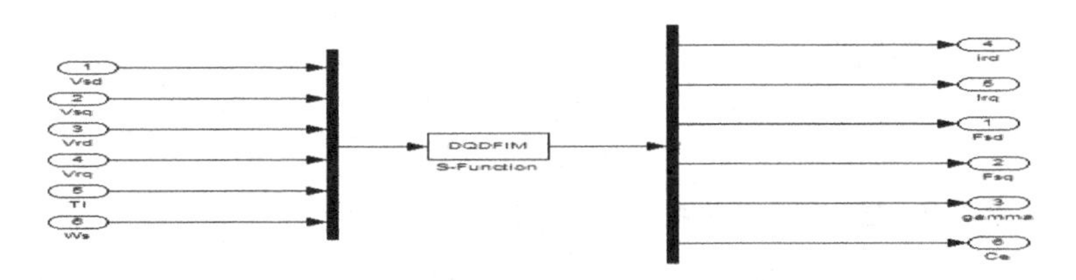

Figure 3.1 MATLAB/SIMULINK S-function for DFIM open loop

The function DQDFIM is given by:

```
function [sys,x0,str,ts]=DQDFIM(t,x,u,flag)
Rr=3.805;Rs=4.84; M=0.258;Ls=0.274; Lr=0.274;Jm=0.031;
p=2;fm=0.008;
Ts=(Ls/Rs); sigma=1-(M^2/Ls/Lr);gamma=(Rs/sigma/Ls)+
  (Rr*M^2/sigma/Ls/Lr^2);
Tr=Lr/Rr;K=M/sigma/Ls/Lr; a=(1-sigma)/sigma;a1=Rs/(sigma*Ls);
  mm=((p*M)/Ls);
switch flag
    case 0
    sizes=simsizes;
    sizes.NumContStates=5;
    sizes.NumDiscStates=0;
    sizes.NumOutputs=6;
    sizes.NumInputs=6;
    sizes.DirFeedthrough=0;
    sizes.NumSampleTimes=1;
    sys=simsizes(sizes);
    str=[];
    ts=[0 0];
    x0=[0 0 0 0 0];

 case 1
  sys(1)=-gamma*x(1)+(K/Ts)*x(3)+(u(6)-p*x(5))*x(2)
         -K*p*x(5)*x(4)+(1/(Lr*sigma))*u(3)-K*u(1);
  sys(2)=-gamma*x(2)+(K/Ts)*x(4)-(u(6)-p*x(5))*x(1)
         +K*p*x(5)*x(3)+(1/(Lr*sigma))*u(4)-K*u(2);
  sys(3)=(M/Ts)*x(1)-(1/Ts)*x(3)+u(6)*x(4)+u(1);
  sys(4)=(M/Ts)*x(2)-(1/Ts)*x(4)-u(6)*x(3)+u(2);
  sys(5)=((p*M)/(Jm*Ls))*(x(4)*x(1)-x(3)*x(2))-u(5)
        /Jm-(fm/Jm)*x(5);
  case 3
  sys(1)=x(1); % ird current
  sys(2)=x(2); % irq current
  sys(3)=x(3); % fsd flux
  sys(4)=x(4); % fsq flux
  sys(5)=x(5); % mechanical speed
  sys(6)=mm*(-x(3)*x(2)+x(4)*x(1)); % Torque
  case{2,4,9}
        sys=[];
  otherwise
  error(['unhandled flag=',num2str(flag)]);
end
```

3.2.2 Exact feedback linearisation controller

The objective is to control the speed and the stator flux aligned with the d axis; therefore, the output vector is written as:

$$y = \begin{bmatrix} h_2(x) \\ h_4(x) \end{bmatrix} = \begin{bmatrix} \phi_{sd} \\ \Omega \end{bmatrix} \tag{3.7}$$

The following notation is used for the lie derivatives of a function:

$$h(x) : \Re^n \rightarrow \Re \tag{3.8}$$

along a vector field:

$$\begin{aligned} f(x) =& (f_1(x), ..., f_n(x)) \\ L_f h(x) =& \sum_{i=1}^{n} \frac{\partial h}{\partial x_i} f_i(x) \end{aligned} \tag{3.9}$$

Where the derivative of the output along the vector field $f(x)$ is defined iteratively as:

$$L_f h(x) = L_f(L_f^{i-1}) \tag{3.10}$$

Applying the following change of coordinates yields:

$$\begin{bmatrix} \zeta_1 \\ \zeta_2 \\ \zeta_3 \\ \zeta_4 \end{bmatrix} = \begin{bmatrix} \hat{h}_2(x) \\ L_f \hat{h}_2(x) \\ \hat{h}_4(x) \\ L_f \hat{h}_4(x) \end{bmatrix} \tag{3.11}$$

Where the $(\hat{h})$ symbol represents the estimated stator flux and speed . To obtain the control law, Equation (3.11) needs to be differentiated as follows:

$$\begin{bmatrix} \dot{\zeta}_1 \\ \dot{\zeta}_2 \\ \dot{\zeta}_3 \\ \dot{\zeta}_4 \end{bmatrix} = \begin{bmatrix} L_f \hat{h}_2(x) \\ L_f^2 \hat{h}_2 + L_{g_1} L_f \hat{h}_2 u_{rd} + L_{g_2} L_f \hat{h}_2 u_{rq} \\ L_f \hat{h}_4(x) \\ L_f^2 \hat{h}_4 + L_{g_1} L_f \hat{h}_4 u_{rd} + L_{g_2} L_f \hat{h}_2 u_{rq} + \frac{f_m}{J_m^2} \hat{\tau}_l \end{bmatrix} \tag{3.12}$$

Where the number of differentiations represent the relative degree of the system which is defined for both outputs by r_1, r_2 respectively and they are equal to 2, then $r_1 + r_2 = 4$, which is equal to the number of state; therefore, the system admits an exact feedback linearisation . Rewriting Equation (3.12) in matrix form gives:

$$\begin{bmatrix} \dot{\zeta}_1 \\ \dot{\zeta}_2 \\ \dot{\zeta}_3 \\ \dot{\zeta}_4 \end{bmatrix} = \begin{bmatrix} \zeta_2 \\ v_1 \\ \zeta_4 \\ v_2 \end{bmatrix} \tag{3.13}$$

So, in order to construct the control vector Equation (3.14) is written in matrix form as follows:

$$\begin{bmatrix} \dot{\zeta}_2 \\ \dot{\zeta}_4 \end{bmatrix} = \begin{bmatrix} L_f^2\hat{h}_1 \\ L_f^2\hat{h}_2 + \frac{f_m}{J_m^2}\hat{\tau}_l \end{bmatrix} + D(x)\begin{bmatrix} u_{rd} \\ u_{rq} \end{bmatrix} \tag{3.14}$$

The decoupling matrix $D(x)$ is defined as follows:

$$D(x) = \begin{bmatrix} L_{g_1}L_f\hat{h}_2 & 0 \\ 0 & L_{g_2}L_f\hat{h}_4 \end{bmatrix} \tag{3.15}$$

Where

$$\begin{aligned} L_{g_1}L_f\hat{h}_2 &= \frac{M}{L_rT_s\sigma} \\ L_{g_2}L_f\hat{h}_4 &= -\frac{pM}{J_mL_s^2\sigma}\hat{\phi}_{sd} \end{aligned} \tag{3.16}$$

$$\begin{aligned} L_f^2\hat{h}_2(x) = &\frac{\hat{\phi}_{sd}}{T_s^2} - \frac{\hat{u}_{sd}}{T_s} - (\frac{M}{T_s^2} + \gamma\frac{M}{T_s})i_{rd} \\ &-\frac{KM}{T_s^2}(u_{sd} + \hat{\phi}_{sd}) + \frac{M^2}{T_s^2\hat{\phi}_{sd}}i_{rd}^2 \\ &-\frac{M}{T_s^2\hat{\phi}_{sd}}u_{sd}i_{rq} - p\frac{M}{T_s}\Omega i_{rq} \end{aligned} \tag{3.17}$$

$$\begin{aligned} L_f^2\hat{h}_4(x) = &\frac{f_m}{J_m}\frac{L_rf_m\hat{\Omega} + pMi_{rq}\hat{\phi}_{sd}}{L_rJ_m} + \frac{pM^2i_{rq}}{T_sL_rJ_m} \\ &-\frac{pMi_{rd}\hat{\phi}_{sd}}{L_rJ_m}\frac{T_su_{sq} + pT_s\hat{\Omega}\hat{\phi}_{sd} + Mi_{rq}}{T_s\hat{\phi}_{sd}} \\ &+\frac{pM\hat{\phi}_{sd}}{L_rJ_m}(Ku_{sd} + \gamma i_{rq} + Kp\hat{\Omega}\hat{\phi}_{sd}) \\ &-\frac{pMi_{rq}}{j_mL_r}(u_{sd} + \frac{\hat{\phi}_{sd}}{T_s}) \end{aligned} \tag{3.18}$$

The matrix $D(x)$ is non-singular,since its determinant has never been zero for any value of the statoric flux ϕ_{sd}. Therefore, the control vector law $[u_{rd} \quad u_{rq}]^T$ is drawn from Equation (3.14):

$$\begin{bmatrix} u_{rd} \\ u_{rq} \end{bmatrix} = D(x)^{-1}\begin{bmatrix} -L_f^2\hat{h}_1(x) + v_1 \\ -L_f^2\hat{h}_2(x) - \frac{f_m}{J_m^2}\hat{\tau}_l + v_2 \end{bmatrix} \tag{3.19}$$

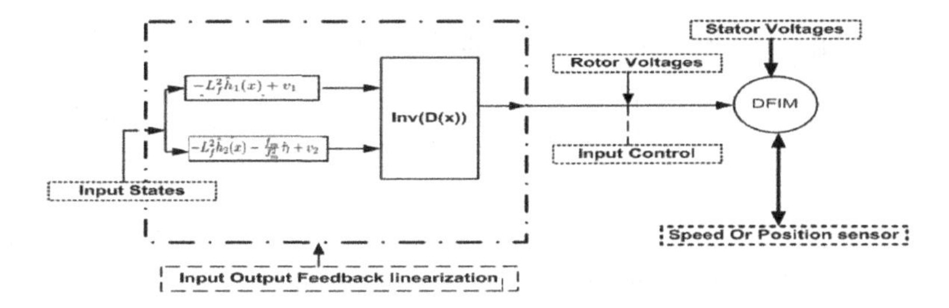

Figure 3.2 The Feedback linearisation Controller

Where v_1, v_2 are the new input vectors that are going to be calculated. It is seen from Equations (3.14), (3.19) that the problem of controlling speed and stator flux is rendered to control a double integrator for the rotor flux loop as well as for the speed loop as shown in Fig 3.3 In order to track the reference trajectories h_2 and

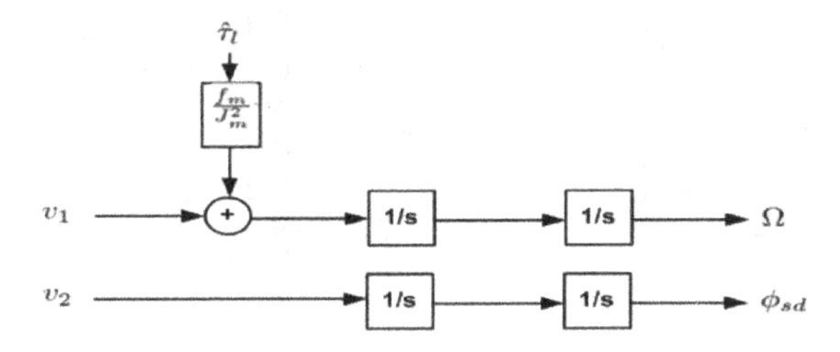

Figure 3.3 The linearised system

h_4, the variations v_1 and v_2 are calculated as follows:

$$\begin{aligned} v_1 =& \ddot{h}_{2ref} - k_{d1}(\dot{h}_2 - \dot{h}_{2ref}) - k_{p1}(h_2 - h_{2ref}) \\ v_2 =& \ddot{h}_{4ref} - k_{d2}(\dot{h}_4 - \dot{h}_{4ref}) - k_{p2}(h_4 - h_{4ref}) \end{aligned} \tag{3.20}$$

By substituting the variations v_1 and v_2 equality found in Equation (3.20) into Equation (3.19); the tracking error equations are given by:

$$\dot{e}_1 = \begin{bmatrix} \dot{e}_{11} \\ \dot{e}_{12} \end{bmatrix} = \begin{bmatrix} \dot{h}_2 - \dot{h}_{r2} \\ \ddot{h}_2 - \ddot{h}_{r2} \end{bmatrix} = \begin{bmatrix} e_{12} \\ -k_{p1}e_{11} - k_{d1}e_{12} \end{bmatrix} \tag{3.21}$$

$$\dot{e}_2 = \begin{bmatrix} \dot{e}_{21} \\ \dot{e}_{2}2 \end{bmatrix} = \begin{bmatrix} \dot{h}_4 - \dot{h}_{r4} \\ \ddot{h}_4 - \ddot{h}_{r4} \end{bmatrix} = \begin{bmatrix} e_{22} \\ -k_{p2}e_{21} - k_{d2}e_{22} \end{bmatrix} \tag{3.22}$$

Writing Equations (3.21) and (3.22) into a matrix form as:

$$\dot{e}_1 = -\begin{bmatrix} 0 & -1 \\ k_{p1} & k_{d1} \end{bmatrix} \begin{bmatrix} e_{12} \\ e_{12} \end{bmatrix} = -A_{ec1}e_1 \qquad (3.23)$$

$$\dot{e}_2 = -\begin{bmatrix} 0 & -1 \\ k_{p2} & k_{d2} \end{bmatrix} \begin{bmatrix} e_{21} \\ e_{22} \end{bmatrix} = -A_{ec2}e_2$$

Where the positive constants k_{p1}, k_{p2}, k_{d1} *and* k_{d2} are obtained so that A_{ec1} and A_{ec2} are Hurwitz and positive definite matrices.

Exact Feedback linearisation controller Matlab/Simulink

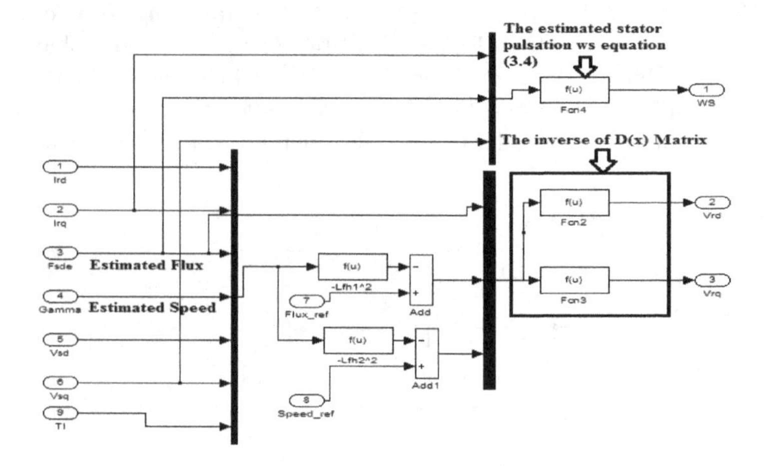

Figure 3.4 MATLAB/SIMULINK block for nonlinear controller

3.2.3 The Luenberger observer for load torque and mechanical speed

Since the load torque is unknown and should be measured using sensors to compensate it, and since sensors are more expensive and therefore a good solution could be the use of the load torque estimator; hence, a second order Luenberger observer is proposed with the stator currents as measured states, the speed is estimated by classical method. From Equation (3.25) we have that the electromagnetic torque and speed equations are simplified to:

$$C_{em} = K_{Te}i_{rq} \qquad (3.24)$$

$$\dot{\hat{\Omega}} = K_{Te} i_{rq} - \frac{f_m}{J_m}\Omega - \frac{\hat{\tau}_d}{J_m} \tag{3.25}$$

Where the parameter K_{Te} represents the torque constant and is defined by :

$$K_{Te} = \frac{pM}{J_m L_s}\hat{\phi}_{sd} \tag{3.26}$$

The estimated stator flux in the field oriented control is given by:

$$\frac{\dot{\hat{\phi}}_{sd}}{T_s} = -\frac{\hat{\phi}_{sd}}{T_s} + \frac{M}{T_s} i_{rq} + u_{sd} \tag{3.27}$$

To estimate the stator flux, a low-pass filter in Fig 3.5 is used, thus Equation (3.27) will be:

$$\hat{\phi}_{sd} = \frac{1}{\tau s + 1}(M i_{rq} + u_{sd}) \tag{3.28}$$

Assuming that the load torque is bounded, which means that it

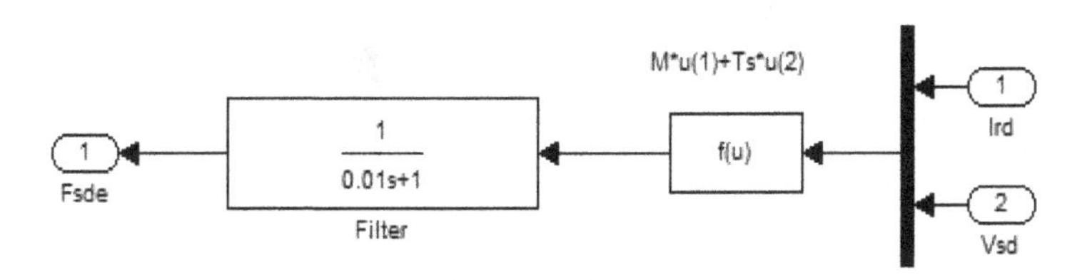

Figure 3.5 Simulink Block for Flux estimation

changes at certain moments, therefore the load torque is defined as:

$$\dot{\tau}_d = 0 \tag{3.29}$$

Equations (3.25)and (3.29) could be written in state equations as follows:

$$\begin{cases} \dot{z} = & Az + B i_{rq} \\ y = & Cz \end{cases} \tag{3.30}$$

Where

$$z = [\Omega \quad \tau_d] \tag{3.31}$$

$$z = \begin{bmatrix} -\frac{f_m}{J_m} & -\frac{1}{J_m} \\ 0 & 0 \end{bmatrix}. \tag{3.32}$$

$$B = [K_{Te} \quad 0] \tag{3.33}$$

The output system y is considered to be the speed such that :

$$C = [1 \quad 0]^T \tag{3.34}$$

Then the Luenberger observer that estimates the speed and the load torque is written as follows :

$$\begin{aligned}\dot{\hat{z}} &= A\hat{z} + Bi_{rq} + H(Cz - C\hat{z}) \\ &= (A - HC)\hat{z} + Bi_{rq} + HCz\end{aligned} \tag{3.35}$$

The differential equation describing the observer error is obtained from Equations (3.30) and (3.36) as:

$$\dot{e} = (A - HC)e = A_h e \tag{3.36}$$

Where the symbol $\hat{()}$ denotes the estimated values, $H = [H_1 \quad H_2]$ is the observer gain matrix which has to be chosen adequately such that the error between the estimated and real states converges to zero as t tends to infinity. The parameters H are chosen so that A_h

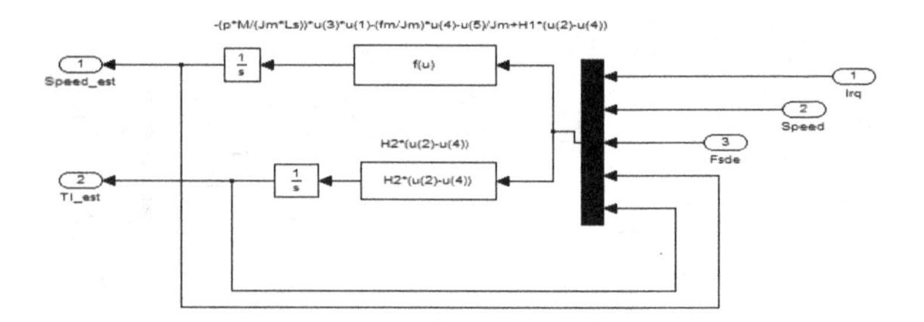

Figure 3.6 MATLAB/SIMULINK block for Luenberger observer

is Hurwitz, therefore Equation (3.36) will be :

$$det(sI - A_h) = 0 \Rightarrow s^2 + (\frac{f_m}{J_m} + H_1)s - \frac{H_2}{J_m} = 0 \tag{3.37}$$

Where s is the Laplace operator and I is the identity matrix; thus by identification to a second order characteristic equation, the parameters H_1 and H_2 are written as:

$$\begin{cases} H_1 = 2\zeta\omega_n \frac{f_m}{J_m} \\ H_2 = -J_m\omega_n^2 \end{cases} \tag{3.38}$$

The parameters ζ and ω_n are chosen to meet the specific requirements

3.2.4 Stability analysis and controller observer convergence

The stability proof of the error convergence of the nonlinear controller and the proposed adaptive neural network observer, using the following Lyapunov function candidate:

$$V = V_1 + V_2 \tag{3.39}$$

Where the functions V_1 *and* V_2 represents the Lyapunov equations of nonlinear controller and the Luenberger observer, respectively, which are given by:

$$V_1 = \frac{e_1^1}{2} + \frac{e_2^2}{2} \tag{3.40}$$

$$V_2 = \frac{e^2}{2} \tag{3.41}$$

Hence the derivative of V is:

$$\dot{V} = e_1\dot{e}_1 + e_2\dot{e}_2 + e\dot{e} \tag{3.42}$$

By substituting Equations (3.36) and (3.23) into equation (3.42) yields:

$$\begin{aligned}\dot{V} =& - A_{ec1}e_1e_1 - A_{ec2}e_2e_2\\ &-A_h e^2\end{aligned} \tag{3.43}$$

After simplification of Equation (3.43) gives:

$$\begin{aligned}\dot{V} =& - A_{ec1}e_1^2 - A_{ec2}e_2^2 - A_e^2\\ \leq& 0\end{aligned} \tag{3.44}$$

V is then a Lyapunov function for the overall system. Consequently, the whole process is stable and the convergence is exponential.

3.2.5 Nonlinear control performance

In Figure 3.7 we note a very good tracking of the mechanical speed as well as the stator flux which is controlled with a very fast response in both transient and permanent regimes since it is unaffected by the change of speed as well as load torque. The electromagnetic torque follows the load torque at a constant speed, where a difference of $\pm 5N.m$ appears between them.

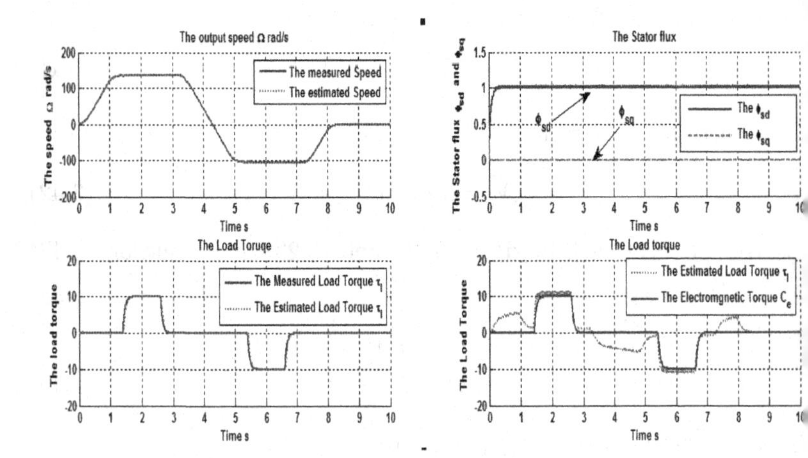

Figure 3.7 The Output Simulation Results for Nonlinear Controller

APPENDIX A

MATLAB/SIMULINK Functions

A.1 MATLAB/SIMULINK BLOCK OF DFIM

MATLAB S-function code given below describes the dynamical model of DFIM in the dq frame of reference where the state vector $x = [x(1), x(2), x(3), x(4), x(5)] = [i_{sd}, i_{sq}, \phi_{rd}, \phi_{rq}, \Omega]$

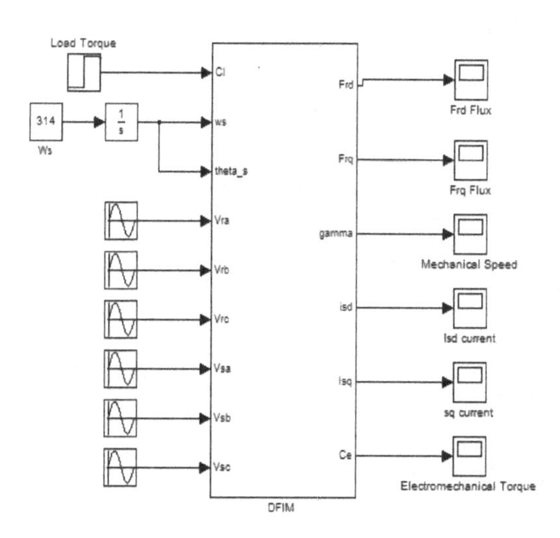

Figure A.1 DFIM open-loop SIMULINK block

A.1.1 MATLAB S-function for the DFIM

```
function [sys,x0,str,ts]=DQDFIM_FRIS(t,x,u,flag,Rr,Rs,M,
 Ls,Lr,Jm,p,fm)
 Rr=3.805;   Jm=0.031; Rs=4.84;
 M=0.258;Ls=0.274; Lr=0.274;Jm=0.031;p=2;fm=0.008;
Ts=(Ls/Rs); sigma=1-(M^2/Ls/Lr);
gamma=(Rs/sigma/Ls)+(Rr*M^2/sigma/Ls/Lr^2);
Tr=Lr/Rr;K=M/sigma/Ls/Lr;
 a=(1-sigma)/sigma;a1=Rs/(sigma*Ls);
a2=Rr/(sigma*Lr);a3=(M*Rr)/(sigma*Ls*Lr);
 a4=(M*Rs)/(sigma*Ls*Lr);a5=M/(sigma*Ls);
a6=M/(sigma*Ls);b1=1/(sigma*Ls);
b2=1/(sigma*Lr);b3=M/(sigma*Ls*Lr);m1=(p^2*M)/Jm;
m2=fm/Jm;m3=p/Jm; m4=m1/m3; m5=m2/m3; mm=((p*M)/Ls);
switch flag
    case 0
    sizes=simsizes;
    sizes.NumContStates=5;
    sizes.NumDiscStates=0;
    sizes.NumOutputs=6;
    sizes.NumInputs=6;
    sizes.DirFeedthrough=0;
    sizes.NumSampleTimes=1;
    sys=simsizes(sizes);
    str=[];
    ts=[0 0];
    x0=[0 0 0 0 0];

 case 1
  sys(1)=-gamma*x(1)+(K/Tr)*x(3)-u(6)*x(2)
        -K*p*x(5)*x(4)+(1/(Ls*sigma))*u(1)-K*u(3);
  sys(2)=-gamma*x(2)+(K/Tr)*x(4)+u(6)*x(1)
        +K*p*x(5)*x(3)+(1/(Ls*sigma))*u(2)-K*u(4);
  sys(3)=(M/Tr)*x(1)-(1/Tr)*x(3)+(u(6)-p*x(5))*x(4)+u(3);
  sys(4)=(M/Tr)*x(2)-(1/Tr)*x(4)-(u(6)-p*x(5))*x(3)+u(4);
  sys(5)=((p*M)/(Jm*Ls))*(x(3)*x(2)-x(1)*x(4))-u(5)/
       Jm-(fm/Jm)*x(5);
  case 3
        sys(1)=x(1);    % isd current
        sys(2)=x(2);    % isq current
        sys(3)=x(3);    % frd flux
        sys(4)=x(4);    % frq flux
        sys(5)=x(5);    % mechanical speed
        sys(6)=mm*(x(3)*x(2)-x(4)*x(1));   % torque
    case{2,4,9}
       sys=[];
   otherwise
           error(['unhandled flag=',num2str(flag)]);
end
```

A.2 MATLAB/SIMULINK BLOCK OF IM

The MATLAB S-function code given below describes the dynamical model of induction machine in an open loop in the $\alpha\beta$ frame of reference, where the state vector $x = [x(1), x(2), x(3), x(4), x(5)] = [i_{s\alpha}, i_{s\beta}, \phi_{r\alpha}, \phi_{r\beta}, \Omega]$.

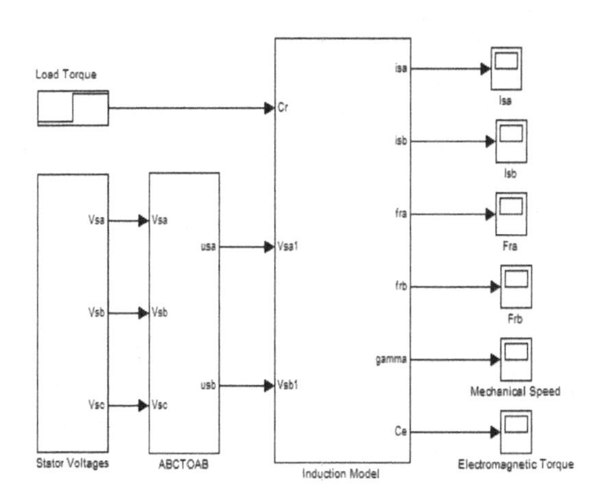

Figure A.2 Induction motor open-loop SIMULINK block

A.2.1 MATLAB S-function for the IM

```
function [sys,x0,str,ts]=IM_IFAB(t,x,u,flag,Rr,Rs,M,Ls,
  Lr,Jm,p,fm)
Rr=3.805;Rs=4.84;
M=0.258;Ls=0.274;
Lr=0.274;Jm=0.031;p=2;fm=0.00114;
Ts=(Ls/Rs);
sigma=1-(M^2/Ls/Lr);gamma=(Rs/sigma/Ls)+(Rr*M^2/sigma/
  Ls/Lr^2);
Tr=Lr/Rr;K=M/sigma/Ls/Lr;a=(1-sigma)/sigma;a1=Rs/
  (sigma*Ls);
a2=Rr/(sigma*Lr);a3=(M*Rr)/(sigma*Ls*Lr);a4=(M*Rs)/
  (sigma*Ls*Lr);
a5=M/(sigma*Ls);a6=M/(sigma*Ls);b1=1/(sigma*Ls);
b2=1/(sigma*Lr);b3=M/(sigma*Ls*Lr);m1=(p^2*M)/Jm;
m2=fm/Jm;m3=p/Jm;
m4=m1/m3;
```

```
m5=m2/m3;
mm=((p*M)/Ls);
switch flag
    case 0
    sizes=simsizes;
    sizes.NumContStates=5;
    sizes.NumDiscStates=0;
    sizes.NumOutputs=6;
    sizes.NumInputs=3;
    sizes.DirFeedthrough=0;
    sizes.NumSampleTimes=1;
    sys=simsizes(sizes);
    str=[];
    ts=[0 0];
    x0=[0.5 0.5 0.5 -0.5 0.1];

     case 1
         AA=[Rs 0 0 0 0;
             0 Rs 0 0 0;
             -(M/Tr) 0 1/Tr p*x(5) 0;
             0 -(M/Tr) -p*x(5) 1/Tr 0;
             -mm*x(4) mm*x(3) 0 0 -fm];
         BB=[Ls*sigma 0 (M/Lr) 0 0;
             0 Ls*sigma 0 (M/Lr) 0;
             0 0 1 0 0;
             0 0 0 1 0;
             0 0 0 0 -Jm];
         U=[u(1);u(2);0;0;u(3)];
         II=[x(1);x(2);x(3);x(4);x(5)];
         sys=-inv(BB)*AA*II+inv(BB)*U;
     case 3
         sys(1)=x(1);                                % isa current
         sys(2)=x(2);                                % isb current
         sys(3)=x(3);                                % fra flux
         sys(4)=x(4);                                % frb flux
         sys(5)=x(5);                                % Mechanical spe
         sys(6)=(3/2)*(x(3)*x(2)-x(4)*x(1));         % Ce torque
        case{2,4,9}
        sys=[];
    otherwise
           error(['unhandled flag=',num2str(flag)]);
end
```

A.3 THE INPUT VOLTAGES

The stator and rotor input vector voltages are given by the following equations:

A.3.1 Stator input voltages equations

$$
\begin{aligned}
V_{sa} &= 220\sin(2\pi f_s t) \\
V_{sb} &= 220\sin(2\pi f_s t - \frac{2\pi}{3}) \\
V_{sa} &= 220\sin(2\pi f_s t - \frac{4\pi}{3})
\end{aligned}
\tag{A.1}
$$

A.3.2 Rotor input voltages equations

$$
\begin{aligned}
V_{ra} &= 12\sin(2\pi f_s t) \\
V_{rb} &= 12\sin(2\pi f_s t - \frac{2\pi}{3}) \\
V_{ra} &= 12\sin(2\pi f_s t - \frac{4\pi}{3})
\end{aligned}
\tag{A.2}
$$

Where $fs = 100\pi$ is the frequency of the input vectors; the three-phase input vectors are addressed to a two-phase DFIM using the Park transformation as shown in Figure (A.3)

A.3.3 Clarke and Park transformation SIMULINK model

Clarke and Park transformations are used in vector control (filed-oriented control) of AC machines. The Park transform is used to convert $\alpha\beta$ signals to dq rotating reference frame. Clarke transform is used to convert abc signals to $\alpha\beta$ stationary frame.

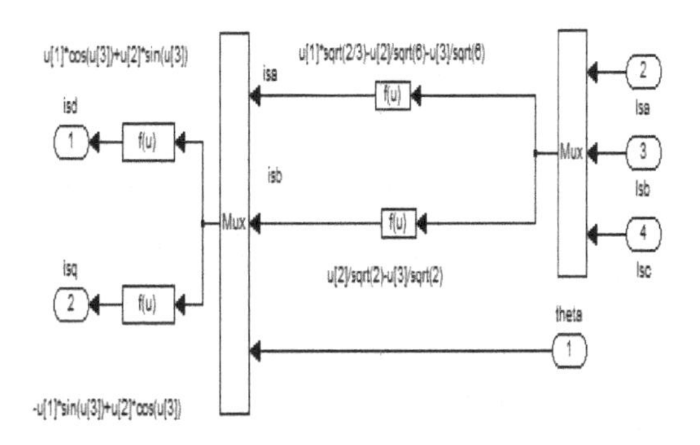

Figure A.3 Three-to-two-phase transformation block

Therefore, the Clarke equation that transforms the three-phase *abc* to $\alpha\beta$ is given by:

$$\begin{bmatrix} V_{s\alpha} \\ V_{s\beta} \end{bmatrix} = \begin{bmatrix} \sqrt{\frac{2}{3}} & -\frac{1}{\sqrt{6}} & -\frac{1}{\sqrt{6}} \\ 0 & \frac{1}{\sqrt{2}} & -\frac{1}{\sqrt{2}} \end{bmatrix} \begin{bmatrix} V_{sa} \\ V_{sb} \\ V_{sc} \end{bmatrix} \tag{A.3}$$

The Park transformation equation from $\alpha\beta$ to dq is given by:

$$\begin{bmatrix} V_{sd} \\ V_{sq} \end{bmatrix} = \begin{bmatrix} \cos(\omega_s) & \sin(\omega_s) \\ -\sin(\omega_s) & \cos(\omega_s) \end{bmatrix} \begin{bmatrix} V_{s\alpha} \\ V_{s\beta} \end{bmatrix} \tag{A.4}$$

APPENDIX B

Induction Machine Parameters

Table B.1 Parameters of the DFIM

Designation	Parameter	Value
Rotor resistance	R_r	3.805 Ω
Stator resistance	R_s	4.85 Ω
Mutual inductance	M	0.258 H
Stator cyclic inductance	L_s	0.247 H
Rotor cyclic inductance	L_r	0.247 H
Rotor inertia	J_m	0.031 Kg/m^3
Pole pair	p	2
Viscous friction coefficient	f_m	0.008 N.m.s/rd
Mechanical power	P_m	15 KW
Nominal Stator Voltage	V_s	220 V
Nominal Rotor Voltage	V_r	12 V
Nominal Stator Current	I_s	3.46 A
Nominal Rotor Current	I_r	6.31 A
Nominal speed	Ω_n	1500 rev/min

APPENDIX C

FIELD-ORIENTED CONTROL Controller Parameter Tuning

The parameters of the four controllers shown in Figure 2.6 are given by:

C.1 FIELD ORIENTED CONTROL PARAMETER OF CHAPTER 2

```
clear all;clc
Rr=3.805;Rs=4.84;
M=0.258;Ls=0.274;
Lr=0.274;Jm=0.031;p=2;fm=0.008;Ts=Ls/Rs;
sigma=1-(M^2/Ls/Lr);
gamma=(Rs/sigma/Ls)+(Rr*M^2/sigma/Ls/Lr^2);
Tr=Lr/Rr;K=M/sigma/Ls/Lr;
a=(1-sigma)/sigma;a1=Rs/(sigma*Ls)
;a2=Rr/(sigma*Lr);
a3=(M*Rr)/(sigma*Ls*Lr);
a4=(M*Rs)/(sigma*Ls*Lr);
a5=M/(sigma*Ls);a6=M/(sigma*Ls);
b1=1/(sigma*Ls);b2=1/(sigma*Lr);
b3=M/(sigma*Ls*Lr);m1=(p^2*M)/Jm;
m2=fm/Jm;m3=p/Jm;
```

```
m4=m1/m3;
m5=m2/m3;
gm=Rr+((M^2/Lr/Ts));
mm=(p*M)/Lr;
KI=(p*M)/(Jm*Lr*Ls*sigma);
KK=(p*M)/(Jm*Lr);
K1=Rr+((M^2)/(Ls*Ts));
K2=((sigma*Ls*Ts*Lr)/(Rr*Ls*Ts+M^2));
%---------------PI3---------------------%
zeta=1;
Kt=(p*M)/(Lr);
%wn=4/(zeta*0.06)%4/(zeta*0.015);
%wnf=4/0.04;
%tau=0.01;
wn=4/0.06;
Kp=(2*zeta*wn*Jm-fm)/Kt;
Ki=Jm*wn^2/Kt;
%-----PI1 and PI4-----------------------%
n=10;
h=(Ls*sigma)/(n*Rs);
Kpird=(sigma*Ls)/h;
Kiird=Rr/h;
Kpirq=Kpird%(sigma*Lr)/h;%0.3107;
Kiirq=Kiird%gm/h;%80.9625;
%---------------PI2-----------------------%
F=10;
TF=Tr/F;
KIF=1/(M*TF);
KPF=Tr/(M*TF);
%------------------------------
Kpf=KPF;%(2*wnf*zeta*Ts-1)/M
Kif=KPF;%(wnf^2*Ts)/M
```

Table C.1 Field-Oriented Control Controllers of DFIM

PI1 and PI4	PI2	PI3

n=10;
h=(Ls*sigma)/(n*Rs);
Kpird=(sigma*Ls)/h;
Kiird=Rr/h;
Kpirq=Kpird
Kiirq=Kiird
	F=10;
	TF=Tr/F;
	KIF=1/(M*TF);
	KPF=Tr/(M*TF);
		wn=4/0.06;
		Kp=(2*z*wn*Jm-fm)/Kt;
		z=zeta;
		Ki=Jm*wn^2/Kt;

C.2 VECTOR CONTROL USING NONLINEAR CONTROL PARAMETER OF CHAPTER 3

```
% The machine parameters
Rr=3.805;Rs=4.84;
M=0.258;Ls=0.274;
Lr=0.274;Jm=0.031;p=2;fm=0.008;
Ts=(Ls/Rs);sigma=1-(M^2/Ls/Lr);gamma=(Rs/sigma/Ls)+(Rr*M^2/
  sigma/Ls/Lr^2);
```

```
Tr=Lr/Rr;K=M/sigma/Ls/Lr; mm=((p*M)/Ls);

%% Controller Parameters  PD

% Flux parameters
Kpf=4759800;
Kdf=3990

% Mechanical Speed parameters

Kps=899900;

Kds=500;

% Luenberger Observer parameters

wn=50;zeta=0.7;

Jm=0.031;fm=0.008;

H1=2*zeta*wn-(fm/Jm);
H2=-Jm*wn^2;
```

C.3 SIMULINK BLOCK FOR VECTOR CONTROL USING NONLINEAR CONTROLLER

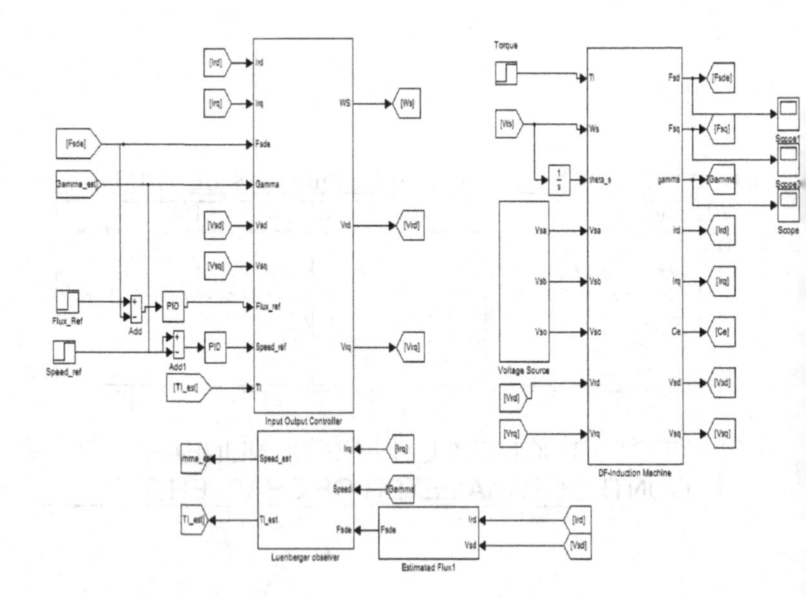

Figure C.1 SIMULINK block for vector control using nonlinear controller

Bibliography

[1] N.P.Quang, J. A. Dittrich (2008): Vector Control of Three-Phase AC Machine *Springer*

[2] Blaschke F (1972) Das Verfahren der Feldorientierung zur Regelung der Asynchronmaschine. Siemens Forschungs- und Entwicklungsberichte. Bd.1, Nr.1/1972

[3] Abolfazl H ,Hossein R, and Amini-Khoei (2015) Sensorless Direct Power Control of Induction Motor Drive Using Artificial Neural Network, *Hindawi Publishing Corporation, Advances in Artificial Neural Systems* Volume 2015, pp.1-9

[4] Andreas M (2011) Recent advances in robust control-Novel approaches and design methods . *InTech.* Croatia

[5] Hasse K (1969) Zur Dynamik drehzahlgeregelter Antriebe mit stromrichtergespeisten Asynchron-Kurzschlulufermaschinen. Dissertation, TH Darmstadt

[6] Quang NP, Dittrich A, and Thieme A (1997) Doubly-fed induction machine as generator: control algorithms with decoupling of torque and power factor. Electrical Engineering / Archiv fr Elektrotechnik, 10.1997, pp. 325-335

[7] Brandstetter P, Sensorless Control of DC Drive Using Artificial Neural Network, *Acta Polytechnica Hungarica* Vol. 11, No 10, pp.5-20

[8] Blaschke F (1972) The principle of field orientation as applied to the new transvector closed loop control system for rotating field machines. *Siemens Rev*, Vol.39, No.5, pp.217-220.

[9] Chen F and Dunnigan M (2004) A new non-linear sliding-mode torque and flux control method for an induction machine incorporating a sliding-mode flux observer. *International Journal of Robust and Nonlinear Control*, Vol.14, No.5, pp. 463-486.

[10] Chiasson J (1996) Nonlinear controllers for an induction motor *Control engineering practice* , Vol 4 No. 7, pp. 977-990

[11] Cherifi D, Miloud Y and Tahri A, (2012) Simultaneous Estimation of Rotor Speed and Stator Resistance in Sensorless Indirect Vector Control of Induction Motor Drives Using a Luenberger Observer . *IJCSI Internation al Journal Of Computer Science Issues*, Vol. 9, Issue 3, No 2, p.325

[12] Daniele B, Antonella F and Matteo R, (2011) Sliding Mode observers for sensorless control of current-fed induction motors, *American Control Conference on OFarrell Street, San Francisco, CA, USA June 29-July 01* pp.763-768

[13] Farrokh, Hashemnia N, and Kahiha A (2010) Robust speed sensorless control of doubly fed induction motor induction machine based on input output feedback lin-earization control using a Sliding mode observor *World Applied Sciences Journal* Vol 10, No.11,pp.1392-400

[14] Iqbal A, Rizwan Kh,(2010) Sensorless Control of a Vector Controlled Three-Phase Induction Motor Drive using Artificial Neural Network, *International Conference on Power Electronics, Drives and Energy Systems(PEDES)Power India* 20-23-Dec 2010, pp.1-5

[15] Khalil N, Abed K, Benalla H (2008) Sensorless direct torque control of brush less ac machine using luenberger observer *Journal of theoretical and applied information technology* Vol.04, No.08, pp. 725-730

[16] Lei-Po L, Zhu-Mu F,Xiao-na S (2012) Sliding mode control with disturbance observer for a class of nonlinear systems *International journal of Automation and Computing*,Vol.9, Issue 5, pp.487-491

[17] Mechrnene A, Zerikat M, and Chekroun S (2012) Adaptive Speed Observer using Artificial Neural Network for Sensorless Vector Control of Induction Motor Drive, *AUTOMATIKA*, Vol. 53, No. 3, pp. 263-271.

[18] Merabet A (2012) Frontiers of model predictive control. edited by Tao Zheng *INTECH* ,pp 109-130.

[19] Marino R., Peresada S. and Valigi P (1993) Adaptive input output linearizing control of induction motors *IEEE Trans. Automat. Control.*, Vol .38, No.2, pp.208-221.

[20] Oscar B, Alkorta P, Jose Maria G and Enrique K (2012) A robust position Control for induction motors using a load torque Observer. *20th Mediterranean Conference on Control Automation (MED)*, Barcelona, Spain, 3-6 July 2012, pp.278-283

[21] Orlowska Kowalska T and Dybkowski M (2012) Performance analysis of the sensorless adaptive sliding- mode neuro-fuzzy control of the induction motor drive with MRAS-type speed estimator *Bulletin of the Polish academy of sciences technical sciences*, Vol. 60, No.1, pp.61-70.

[22] Serhoud H and Benttous D (2012) Sliding mode control of brush-less doubly fed induction machine used in wind energy conversion system *Revue des energies renouvelables* Vol.15 No.2, pp.305-320.

[23] Tarek B, Omari A,(2011) Improved Adaptive Flxu Observer of an Induction Motor with Stator Resistance Adaptation, *Przeglad Elektrotechniczny (Electrical Review)*, Vol. 87, No.09, pp.325-329.

[24] Cherifi Djamila, Miloud Yahia and Tahri Ali, (2012) Simultaneous Estimation of Rotor Speed and Stator Resistance in Sensorless Indirect Vector Control of Induction Motor Drives Using a Luenberger Observer . *IJCSI Internation Journal al Of Computer Science Issues, Vol. 9, Issue 3, No 2*

[25] Daniele Bullo, Antonella Ferrara and Matteo Rubagotti, (2011) Sliding Mode observers for sensorless control of current-fed induction motors, *2011 American Control Conference on OFarrell Street, San Francisco, CA, USA June 29-July 01.*

[26] Seguier G and Notelet F. (2010), Electrotechnique Industrielle, 2eme Edition, Edition Technique et Documentation, Paris, France, 1994.

[27] Pavel Brandstetter, Sensorless Control of DC Drive Using Artificial Neural Network, *Acta Polytechnica Hungarica Vol. 11, No 10.*

[28] Iqbal Arif, Khan Rizwan, (2010) Sensorless Control of a Vector Controlled Three-Phase Induction Motor Drive using Artificial Neural Network, *Power Electronics, Drives and Energy Systems (PEDES)* Power India.

[29] Boufadene, M. (2018). Nonlinear Control Systems using MATLAB. *Boca Raton: CRC Press.*

[30] Boufadene, M., Belkheiri, M. and Rabhi A. (2048): Adaptive nonlinear observer augmented by radial basis neural network for a nonlinear sensorless control of an induction machine, *Int. J. Automation and Control, Vol. 12, No. 1, pp.2743.*

Index

Printed in the United States
by Baker & Taylor Publisher Services